Super-Fun™
Problem-Solving Cards

101 SNAP-APART STORY PROBLEMS THAT BUILD ESSENTIAL MATH SKILLS

BY LYNN BEEBE

ILLUSTRATIONS BY PETER GEORGESON

SCHOLASTIC
PROFESSIONAL BOOKS

NEW YORK • TORONTO • LONDON • AUCKLAND • SYDNEY

Super-Fun Math Problem-Solving Cards copyright © 1998 by Lynn Beebe.
Illustrations copyright © 1998 by Peter Georgeson.

These story problems may be reproduced for classroom use. No other part of the publication may be reproduced
in whole or in part, or stored in a retrieval system, or transmitted in any forms or by any means, electronic, mechanical,
photocopying, recording, or otherwise, without permission of the publisher. For information, regarding permission,
write to Scholastic Professional Books, 555 Broadway, New York, NY 10012.

ISBN: 0-590-25541-X

Illustrations by Peter Georgeson
Book design by Amy Redmond
Art direction by Simon Sung
Production by Amy Sinclair

Super-Fun Math is produced by **becker&mayer!**, Kirkland, Washington.

Welcome to the world of

Super-Fun Math

Problem-Solving Cards!

> "Problem solving should be the central focus of the math curriculum."
> —*the National Council of Teachers of Mathematics*

The cards in this book are designed to bring a light-hearted touch to the primary goal of the NCTM Standards—teaching math problem solving. These funny, engaging story problems are sure to encourage even your most reluctant learners to think critically and creatively about the math encountered in everyday situations—from divvying up a pizza pie to calculating the cost of buying supplies for a pet pig.

These cards will fit easily into any math program because the problems are based on the NCTM Standards for elementary students. The specific standards covered are:

#1 Problem Solving	#8 Whole Number Computations	#11 Statistics and Probability
#3 Reasoning	#9 Geometry and Spatial Sense	#12 Fractions and Decimals
#5 Estimation	#10 Measurement	#13 Patterns and Relationships

Organization of This Book

Problem solving and reasoning, the two all-important NCTM Standards, are inherent in each of the cards. The other six content standards have been grouped into separate sections, enabling you to easily access the problems that match your needs and curriculum focus.

In some cases, you'll find problems that involve more than one standard; we have placed these under the most appropriate heading. The cards have been sequenced from easiest to most difficult—the larger the number, the more challenging the math task. Each card has the problem's solution on its back, so children can check their own answers.

How to Use These Cards

Matching the Cards with Kids: The great thing about *Super-Fun Math Problem-Solving Cards* is that they help you help your students in so many ways! Use them to:

- motivate kids who don't gravitate to math—they won't be able to resist these humorous story problems!;

- give individual children practice with skills and concepts they find particularly difficult;

- provide entertaining math challenges for advanced students who are ready for independent problem solving;

- help your class see the real applications for each of the problem-solving strategies.

Learning Center Fun: Place the cards at an independent station. Then, encourage kids to visit, work at their own pace, and record their solutions in a special section of their math notebooks. Students can correct the problems themselves. After all, reading the answers is half the fun!

Problem of the Day: Each morning, select one card to be the Problem of the Day. Copy the story problem on the chalkboard for kids to solve in their spare time. Later in the day, invite kids to share their answers. Did everyone use the same strategy?

Group Learning: Divide your class into cooperative groups. Distribute a card to each one, challenging each team to work collaboratively to solve the problem in 10 minutes or less.

More Helpful Hints

Maintaining Your Cards: Once removed from the book, these cards can be stored in a recipe box (with dividers for each content standard), in labeled manila envelopes, or in handy pocket folders. Laminating or covering your cards with clear contact paper will ensure their long life.

Empowering Young Problem Solvers: A general introduction to the problem-solving strategies— making a diagram, working backwards, looking for a pattern—will boost your students' confidence and enable them to use these cards independently and effectively.

Using Calculators: Some of these problems lend themselves to the use of calculators. Since the primary goal of the NCTM Standards is to teach kids to become accomplished problem solvers, you might consider letting them use calculators for some of the harder computations.

Extending Learning: After your class is familiar with the cards, challenge kids to create their own index-card story problems (with questions on the front and solutions on the back). Number the cards, then add them to your stash for the rest of the class to solve!

Fuzzy Purple Gorillas!

Q: You are trying to get a prize from the machine outside the grocery store. From past experience, you know it will take you 5 tries to get what you want: a little purple fuzzy gorilla. Each try costs you a quarter, and you can always get change in the store. You had 3 quarters, 6 nickels, 4 dimes, and 7 pennies in your pocket. Did you have to borrow any money from your mom?

Yikes Hike!

Q: Your class is going on a hike. If you hiked 1.7 miles before you got scratched by blackberry vines, 2.8 miles before you saw a snake, another 1.2 miles before you slipped in the mud, and 0.6 more of a mile until you got to the lake, then how long was the hike to the lake?

Attention Shoppers!

Q: Your mother sent you to the store. She gave you a list: a carton of milk, a bag of a dozen bagels, 4 cans of tuna, a bunch of 9 bananas, and a bag of cat food. Can you go in the rapid checkout line that says "Nine items or less"?

Slug Slime

Q: You have to amuse your cousin from out of town. You decide to show him your banana slug. It crawls onto his arm. Soon he has a trail of slime on his arm. If he tries to get the slime off with soap and water for 25 seconds, then a wash cloth for 30 seconds, and finally with the scrub brush for 1 minute, how long will he spend washing his arm in all?

The Solution

A: No, you needed $1.25. You had $0.75 + $0.30 + $0.40 + $0.07, or $1.52. But now you are wondering if the little fuzzy purple gorilla was really worth it!

The Solution

A: Yes, since a bunch of bananas counts as 1 item, you only have a total of 8 items. Why not treat yourself to a pack of gum while waiting in line!

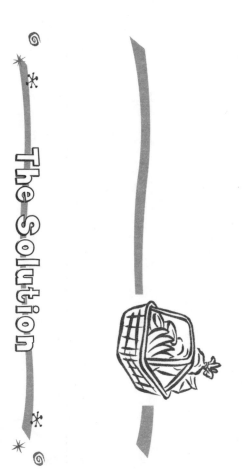

The Solution

A: 6.3 miles. Now, the only problem is you are scratched, muddy, shaken by the snake encounter, and hungry. PLUS you still have to hike back!

The Solution

A: He spent 1 minute and 55 seconds washing. Unfortunately, the slug slime is super hard to get off and now he's gone back into the bathroom... I guess you'll have to explain to Mom why he isn't coming down to dinner!

You Scream for Ice Cream!

Q: You really want an ice cream bar. You break open your piggy bank and ride your bike to the store. If ice cream bars cost about $2.89 a box, and there are only pennies, nickels, and dimes in the jar, what are the fewest coins you can bring (since they have to fit in your pocket) and what are they?

SUPER CHALLENGE: What if you find 3 quarters in the jar? Now what are the fewest coins you can bring?

Lunch Crunch

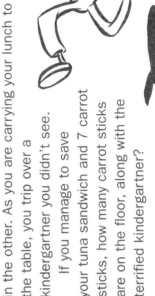

Q: It's lunchtime and your stomach is growling. You decide to get carrot sticks to go with your tuna sandwich. You scoop up 12 carrot sticks in one hand and twice as many in the other. As you are carrying your lunch to the table, you trip over a kindergartner you didn't see.

If you manage to save your tuna sandwich and 7 carrot sticks, how many carrot sticks are on the floor, along with the terrified kindergartner?

Fabulous Frogs

Q: You are trying to catch frogs. You catch 5 green frogs, 2 leopard frogs, and 3 bull frogs. Each time you add 3 more frogs to the bucket, 1 jumps out. How many frogs do you have?

Adding to your Allowance

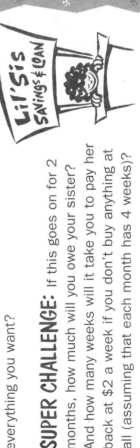

Q: You get $5 a week for your allowance. Every week you spend $1.50 on candy, $1.25 to play video games, $1 for a comic book, $1.45 on pizza, and $0.55 on bus fare. Your little sister never spends any of her allowance. How much money do you have to borrow from your little sister each week to buy everything you want?

SUPER CHALLENGE: If this goes on for 2 months, how much will you owe your sister? And how many weeks will it take you to pay her back at $2 a week if you don't buy anything at all (assuming that each month has 4 weeks)?

The Solution

A: You have to bring a total of 33 coins: 28 dimes, 1 nickel, and 4 pennies. Hopefully your pocket won't burst!

SUPER CHALLENGE: If there are 3 quarters in the jar, you can now bring 28 total coins, or the 3 quarters along with 21 dimes and 4 pennies.

The Solution

A: 8 frogs. You put 3 green frogs in; 1 jumps out. (2 are now in the bucket.) You put in 2 more green frogs and 1 leopard frog; 1 jumps out. (4 are in the bucket.) You put in 2 leopard frogs and 1 bullfrog; 1 jumps out. (6 are in the bucket.) You put in the last 2 bullfrogs—now 8 are in the bucket!

The Solution

A: 29 carrot sticks, and that's if you count really fast. Five seconds later, the whole sixth grade comes into the lunchroom… and the carrot sticks are turned to mush by trampling feet! Luckily, you and the kindergartner escape to safety!

The Solution

A: You have to borrow $0.75 (75 cents) each week.

SUPER CHALLENGE: You will owe her $6. It will take you 3 weeks to pay her back.

A Mouse in the House #12
Multiplication

Q: Your teacher got a whole bunch of mice for your class and he has offered free mice to all of the students. You decide to bring 3 mice home. If you keep them under the bed, you figure that your mom will never find them. At first everything goes well. Then, the mice start having babies! Each mouse that you brought home has 9 babies. How many mice do you have to hide now?

Weighing in #14
Multiplication

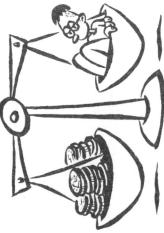

Q: You have just won a contest on TV. Now you have to decide which is the best prize: a bag of quarters that weigh what you weigh or $500. If you know that 5 quarters = 1 ounce and 16 ounce = 1 pound, and you weigh 100 pounds, what is the best choice? You can use a calculator to help you.

Say Cheese! #11
Multiplication

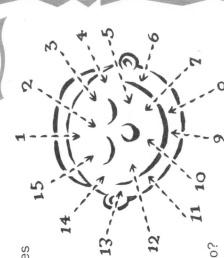

Q: A person uses 15 muscles to smile. Your dad is taking family pictures and keeps telling everyone to smile. There are 5 people in the picture: you, your mom, your kid brother, your Aunt Helen, and Uncle Harry. How many muscles would be used if everyone smiles for the photo?

The Magic Beanstalk #13
Multiplication

Q: You buy a strange bean at a magic store. You plant it when you get home, and it starts to grow the next day. It grows very fast. If it grows 2 feet a day, how tall will the plant be after a week?

The Solution

A: 30 mice (3 grown-up mice and 27 babies). But that was before your mother found them, shouted, "Eeek!" and made you give them up. Now you need to fill the void. On the school bus, a girl said something about giving away her pet boa. Now, who would notice a little old snake in your bedroom...?

The Solution

A: Take the quarters. Your weight would equal 8,000 quarters (16 x 5 = 80 quarters in a pound; 80 x 100 pounds = 8,000 quarters), and 8,000 quarters is $2,000!

The Solution

A: 75 muscles. Your kid brother even laughed, and he normally won't do that unless he's getting tickled!

The Solution

A: 12 feet tall. Since the bean didn't grow the day you bought it, that leaves 6 days in the week and 6 x 2 = 12.

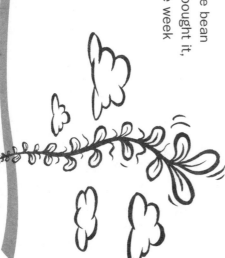

Hoop Dreams

Q: You and a friend are shooting baskets. You tried 15 shots and your friend tried 16. $\frac{1}{3}$ of your shots went in and $\frac{1}{4}$ of your friend's shots went in. If your brother is giving you a nickel for each shot that you and your friend make, how much does he owe you altogether?

Tasty Treats

Q: It's your turn to bring treats for your soccer team. Granola bars come in 1-pound and 2½-pound boxes. It doesn't say on the boxes how many packs are inside of each. You find the 1-pound box partway opened so you can see inside. You count 8 packages inside. You count 8 packages (none had been snitched). Will the bigger 2½-pound box be enough for the team of 18 players? Will there be any leftovers?

Brick by Brick

Q: Your neighbor has a whole pile of bricks to move. She says she will pay you to move them and asks if you want to get paid by the brick (at 3 cents a brick) or by the hour (at $3 an hour). There are about 650 bricks. It takes you 1 minute to move 2 bricks and return for more. Which is the better deal?

Off to See the Stars Play

Q: Your soccer team is raising money for a trip to see the national team play. The total cost of the trip is $2,640. There are 16 players on the team. How much should each player raise?

SUPER CHALLENGE: One team member, Danny, always collects twice as much as the others because he has dozens of relatives he can ask for money. If Danny gathers twice the amount asked, the money needed by the rest of the players will now be lowered to _____?

The Solution

A: Yes, since there will be 20 packs in the bigger box. This means there will not only be enough for the whole team, there will even be 2 left over.

The Solution

The Solution

A: $165.

SUPER CHALLENGE: $154 each. Danny will gather $330. That will leave $2,310 for the rest of the team to earn. Divided by 15 players, that makes $154 each. All right, Danny!

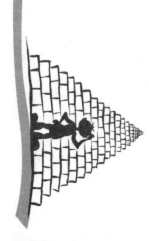

The Solution

A: 45 cents. You made 5 shots and your friend made 4: 4 + 5 = 9 and 9 x 5 cents = 45 cents.

The Solution

A: By the brick, you would earn 6 cents a minute, or $3.60 an hour. Hmm, here comes your little brother...maybe if you split your earnings, he'd do some work for you?

Keep the Candy

Q: For your birthday, your grandmother sent you a box of 48 chocolates. You were planning to eat them all by yourself until you got sick. But your parents have other plans for the candy. They want you to share them with some friends. How many friends can you invite over to share your chocolates evenly and still get at least 8 whole chocolates for yourself?

SUPER CHALLENGE: What if it turns out that your mom had sneaked 8 candies for herself before you even had one? Now how many friends can you invite and still get 8 for yourself?

Piles of Clothes

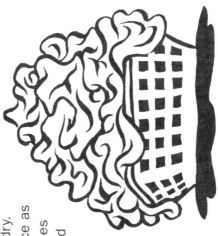

Q: It's your week to fold laundry. In the huge pile, there are twice as many T-shirts as towels, 3 times as many socks as T-shirts, and the same number of pairs of jeans as towels. If there are 10 T-shirts in the pile, how many pairs of jeans, towels, and socks are there?

Do I Look Like a Bank?

Q: You have $20. Your Dad asks to borrow half your money. Then your mom asks to borrow half of what you have left. Then your sister asks to borrow half of what you have after your mom leaves. What are you left with?

Moon Mission

Q: You are having a very strange dream. You are jogging to the moon in a commercial for some new sneakers. The moon is about 239,000 miles from the Earth. Your jogging speed is 10 miles an hour. You are now 10 years old. How old would you be when you arrive? (Remember, there are 24 hours in a day, 7 days in a week, 4 weeks in a month, and 12 months or 365 days in a year.)

The Solution

A: You could have invited over 5 friends and still had 8 chocolates for yourself.

SUPER CHALLENGE: Now you can only invite over 4 friends. (Hey, you better talk to your mom about that chocolate craving of hers!)

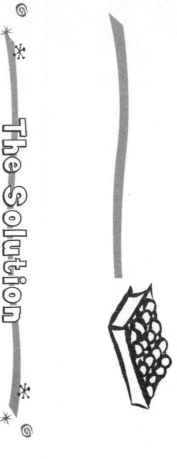

The Solution

A: 5 towels, 5 pairs of jeans, and 30 socks. And guess what? All of the socks have mates. I guess the dryer wasn't very hungry today!

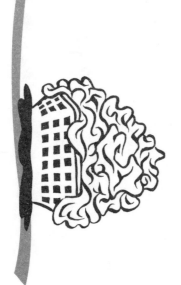

The Solution

A: You have $2.50 left. Uh oh…maybe you can borrow some money from your brother.

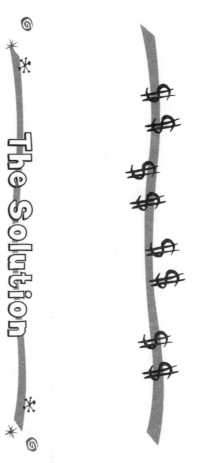

The Solution

A: You would be almost 13 years old!

239,000 miles ÷ 10 miles per hour = 23,900 hours of jogging to get to the moon.

23,900 ÷ 24 hours = 995.8 days to get to the moon.

995.8 days ÷ 365 days in a year = 2.73 years.

10 years + 2.73 years makes you almost 13!

Good thing you woke up, or you would have missed part of your childhood!

Back in Time

Q: You have just ordered lunch at your favorite fast food place, when you are suddenly transported back in time to the year 1940. If most things cost about 10 times less then, what would your $2 shake, $1.50 fries, and $3 hamburger total in 1940?

T-shirts for Sale

Q: Your sister's basketball team is selling T-shirts with their pictures on them to raise money to travel to a tournament. There are 12 girls on the team. The trip will cost $215 per girl. If the girls make $2.50 on each shirt they sell, how many will each girl need to sell to go on the trip?

The Ants Go Marching

Q: You are at a picnic. The ants have found you! You decide to find out how long it takes an ant to walk across your blanket, get a bread crumb, and carry it back to the grass. It takes the first ant 50 seconds, the second ant 45 seconds, the third ant 1 minute, and the fourth ant 53 seconds. What is the average time it takes for an ant to make the round trip?

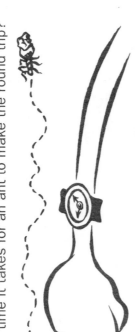

Reward!

Q: You and your two friends are out exploring in the woods behind your house. You find a bicycle. You realize it belongs to your neighbor. He has been looking for it all week. He gives you a reward of $45 for finding the bicycle and another $10 for cleaning it up for him. Can you and your friends divide it evenly? How much will each of you get?

The Solution

A: 65 cents! Now the question is: How do you get back to the future?

The Solution

A: 52 seconds. Did you realize that while you were timing the ants, the yellow jackets carried off your dessert?

The Solution

A: 86 T-shirts. But please! Who wants to buy a T-shirt with your sister's picture on it?!

The Solution

A: No, 55 divided by 3 = $18.33333333 (the 3s go on forever). That means two of you will get $18.33 and that one of you will get $18.34. The friend with the spare penny can throw it in a wishing well and wish for the three of you to find more lost bicycles!

Creepy Crawlies

Q: Your mother asks you to hang up the wet doormat after a big rain storm. When you lift it up you see worms, centipedes, and slugs. There are 4 centipedes. Half of the creatures are slugs. There are twice as many slugs as centipedes. How many creatures were under the mat?

Birthday Math

Q: Your grandfather gives you $100 for your birthday. He says you can spend ½ of it now and ¼ of it next month, then the rest has to go in your saving account. How much will you have to save?

SUPER CHALLENGE:

The first month you bought a snake that cost $50. The next month you bought a heater for its cage that cost $10. How much do you have left to spend?

Insect Dreams

Q: You left your bedroom window open and the light on. There are moths, mosquitoes, and June bugs inside now. You manage to shoo ⅓ of them outside, before you give up and go to sleep. If you put 6 insects out, how many are left?

Surprise!

Q: You are going to surprise your Dad by making him a birthday cake. You've never done this before, but how hard can it be? You want to make a really big cake. You decide to triple the recipe. If the original recipe called for ½ cup of cream, ⅓ cup of butter, and 1 teaspoon of vanilla, how much of each of these would you need now?

The Solution

A: 16 creatures: 8 slugs, 4 centipedes, and 4 worms. You scoop them up and put them in the back yard—that's a better spot for their little tea party!

The Solution

A: You will put $25 into savings.

SUPER CHALLENGE: You still have $15 left to spend.

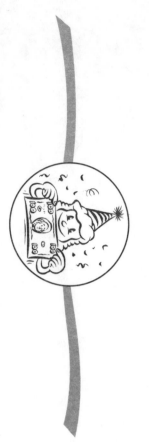

The Solution

A: There are 12 left inside.

The Solution

A: 1½ cups of cream, 1 cup of butter, and 3 teaspoons of vanilla. Unfortunately, you forgot to use a cake pan that was three times bigger. Hey, what's that terrible smell coming from the oven?!

Pizza Party!

Q: You just made your first pizza all by yourself. You take it out of the oven, put it on the table, and cut it into 4 equal slices. The phone rings. You turn to answer it and your pet ferret, who loves pizza, grabs a slice and runs off with it. What fraction of the pizza is left?

SUPER CHALLENGE: How can you cut the remaining 3 pieces of the pizza so that the 4 people in your family will each get an equal share? Hint: Draw a picture to prove your answer.

A Day in Your Life

Q: Your parents think you watch too much TV. You want to convince them that you don't. You decide to record how you spend your time on an average day:

Sleep: 8½ hours	**Homework:** 2 hours
School (with travel time): 8 hours	**Chores:** ½ hour
Meals: 1 hour	**Shower/bath:** ½ hour
Talking with friends: 1 hour	**TV:** 2 hours
	Other: ½ hour

What fraction of your day is spent watching TV?

Strange Friends

Q: You are trying to decide what kind of cupcakes to have for your birthday party. You ask your friends which flavors they like. If ¾ of your friends like chocolate almond mint, ¹²⁄₁₈ of them like vanilla coconut, and ⁷⁄₉ like strawberry watermelon, what flavor should you choose?

Personal Pizzas

Q: You and your friends Beth and Mike have ordered individual-size pizzas. You cut yours into fifths; Mike cuts his into eighths; Beth cuts hers into sixths. If you eat 1 piece while Mike and Beth each eat 2 pieces, who has eaten the most pizza? Who has eaten the second most pizza? Hint: Draw pictures to help you.

The Solution

A: There are 3 out of the 4 pieces of pizza left, or $\frac{3}{4}$ of the pizza.

SUPER CHALLENGE: You would have to cut the remaining 3 pieces into 4 pieces each. Each of these pieces would be $\frac{1}{16}$ of the pizza. Each person in the family would get 3 of these pieces of pizza. Or you could cut each piece in half, give a half to each person, and then cut the last 2 halves in half again and give 1 of those to each.

The Solution

A: $\frac{1}{12}$ of your day. You showed your chart to your parents, and all was going well, until your Dad realized that you had the time to be doing more chores . . . so now you are sorting cans and bottles for recycling instead of watching *Brady Bunch* re-runs.

The Solution

A: Looks like strawberry watermelon is the best choice, since $\frac{7}{9}$ is more than $\frac{3}{4}$ or $\frac{12}{18}$. (The common denominator is 72, so $\frac{48}{72}$ like chocolate almond mint, $\frac{45}{72}$ like vanilla coconut, and $\frac{56}{72}$ like strawberry watermelon.)

P.S. Your friends have very strange tastes in cupcakes!

The Solution

A: Beth has eaten the most. She ate $\frac{2}{6}$ or $\frac{1}{3}$. Mike is next. He ate $\frac{2}{8}$ or $\frac{1}{4}$. You've only finished $\frac{1}{5}$ of yours.

Beth's   Mike's   Your's

Oh, My Tummy Hurts!

Q: You and a couple of friends are having a contest to see who can drink the most milk. (It is a really boring afternoon.) You drink 1½ glasses, take a break, and then drink another glass and a half. One friend drinks 1 glass, and another glass and a half before she stops. Your other friend drinks 2 glasses at once but can't drink any more. Did you win?

Little Lighter Sister

Q: On the Moon, things weigh about ⅙ of what they weigh on Earth. Your little sister weighs 60 pounds on Earth. How much would she weigh if you could actually send her to the Moon?

Eeek!

Q: You're on a camping trip. You and a friend crawl into a tent. You are hoping for a good night's sleep after a long day of hiking in the rain. But once inside the tent, you discover that you're not alone. 4 slugs, 1 spider, and 1 earwig have joined you. What fraction of the creatures in the tent are human?

SUPER CHALLENGE:
What if 2 of the slugs slither out when you come in? Now what fraction of the creatures are human?

Tripping Tortoise

Q: You decide to make an omelet for breakfast. As you are carrying the full carton of a dozen eggs from the refrigerator to the counter, you trip over Speedo, your pet tortoise, and drop the egg carton. If 3 eggs break, what fraction of the eggs are left?

SUPER CHALLENGE:
If you need at least a ½ dozen eggs for your omelet, do you have enough left?

The Solution

A: Yes, you drank a total of 3 glasses compared to your first friend who drank 2½, and your second friend who drank only 2. Too bad you feel too sick to do anything but lie around!

The Solution

A: If everyone and everything had stayed, the fraction of creatures that were human would be ¼.

SUPER CHALLENGE: The human creatures would be ⅓ of the total tent population. (However, your friend is scared silly of spiders so he also made you escort it out, now you humans are still a mere ⅖ of the tent population!)

The Solution

A: Only 10 pounds. Unfortunately, there is no Junior Astronaut Program yet.

The Solution

A: You have 9/12, or ¾ of a dozen left.

SUPER CHALLENGE: Yes, there are 9 eggs left. This is enough for your omelet. Now, if only Speedo would mop up the floor for you!

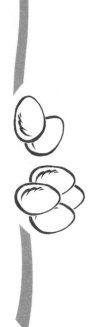

Close Your Mouth, Please!

Q: Your brother no longer has all 20 of his baby teeth. He lost 2 teeth in preschool, then 2 during the summer. While you are walking him to his first day of kindergarten, another tooth falls out. What fraction of his baby teeth are left in his mouth?

Bargain Hunter

Q: You are shopping for school clothes. Everyone is having a sale! In one store, jeans cost $30, but they are 1/2 off the regular price. In another store, the same jeans cost $21, but they are only 1/3 off the regular price. Which store has the best buy, the first or second?

Skunk Time

Q: It is the first day of vacation. You decide that you will sleep for 2/3 of the day, play basketball for 1/8 of the day, eat for 1/24 of the day, and read for 1/12 of the day. How much time will you have left to play with Vanilla, your new pet skunk? (Remember, a day is 24 hours long.)

Fowl Farm

Q: You visit your uncle's farm during the summer. He asks you to feed the ducks and geese. As you watch them eat, you count a total of 54 birds. If 2/9 of your uncle's birds are ducks, how many are geese?

The Solution

A: $^{15}/_{20}$ or $^3/_4$ of his teeth are left. Now if he would only stop showing everyone the latest hole in his mouth and let you get to class!

The Solution

A: The second store, where the jeans cost $14 on sale. At the first store, they cost $15. Since you are such a good shopper, maybe now your mom will let you buy that really cool jacket...

The Solution

A: You have 2 hours to play with Vanilla, after having slept for 16 hours, played basketball for 3 hours, eaten for 1 hour, and read for 2 hours.

The Solution

A: He has 12 ducks and 42 geese. Hey, these feathered fiends must still be hungry, they're eating your shoelaces! Better go and get more food!

Playing With Your Food

Q: There's a bunch of leftover food on the table in the cafeteria. You're feeling creative. In front of you are 16 grapes, 10 raisins, 8 carrot sticks, and 5 banana slices. You decide to make a symmetrical pattern using all of these foods. Draw your pattern on a separate sheet of paper.

Grapes   Raisins   Carrot Sticks   Banana Slices

Amazing Elroy

Q: Elroy, your dog, likes to stop and sniff things when you take him for a walk. But he only stops and sniffs trees, light poles, and fire hydrants. He also only stops at every third thing he passes. (He is a very smart dog.) If you have already passed 2 trees, a fire hydrant, a light pole, and then 3 more trees, will he stop at the fire hydrant just ahead? Draw a diagram to prove your answer.

Copycat

Q: Your brother is imitating you and does everything you do 3 times. You tell him to, "Please stop it," 4 times. Now how many times will your brother say, "Please stop it"?

Call Me!

Q: You just made a new friend at school. You bring home his phone number, but your baby sister starts to chew on the paper. You can still read part of the number, but 2 of the digits are gone. Here is what is left: 76 _ -4 _ 21. You are ready to give in when you remember that he told you there was a pattern in his phone number! What is his number?

SUPER CHALLENGE:
What if the paper read 10 _ -6 _ _ -0?

The Solution

A: No, he would not stop and sniff, since the fire hydrant would only be the second thing since his last sniff stop.

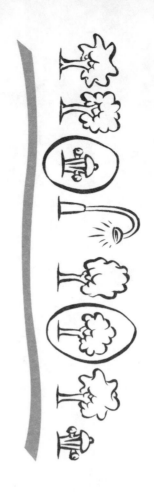

The Solution

A: Answers will vary, but here are a few examples:

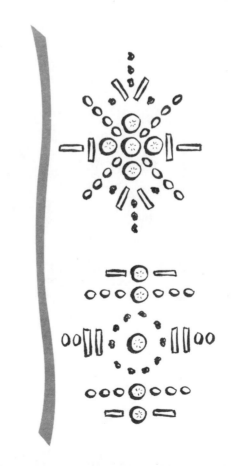

The Solution

A: His number is 765-4321.

SUPER CHALLENGE: 108-6420 (counting backwards by 2s)

The Solution

A: He says it 12 times. After that, he'll no doubt find another way to torture you!

Pea Patterns

Q: You are playing with your food at dinner. You have some peas left. You use 3 peas to make a pyramid shape like this:

This looks cool, so you decide to make bigger pyramid shape using 6 peas. This is even cooler. Next you try to make the next bigger pyramid shape. How many peas do you need?

Bead Work

Q: You are making a necklace for your grandmother. You string 3 red beads, 1 blue bead, 4 green beads, and 1 yellow bead. You repeat this pattern 15 times. How many beads will you need altogether?

Secret Code

Q: You get a message from your friend. It is in code. It says:

13-5-20 1-20 12-21-14-3-8.

Can you break the code?

One Bad Apple

Q: You gathered apples from the tree in your yard to make a pie. The pie takes 6 apples. Every third apple you cut into has worms in it. How many apples should you have gathered in order to have enough with no worms for the pie?

The Solution

A: It would have taken 10 peas. But, even through you tried to explain to your parents that you were in the middle of an important mathematical problem, they made you eat them. Did this ever happen to Einstein?

The Solution

A: It says "meet at lunch." Your friend used 1 for a, 2 for b, 3 for c, and so on.

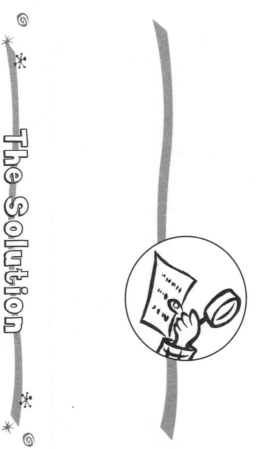

The Solution

A: 135 beads altogether. You will use 45 reds, 15 blues, 60 greens, and 15 yellows. What a lovely work of art it is!

The Solution

A: You need at least 8 apples.

Paper Shredder

Q: A friend passes you a note in class. It says some really embarrassing stuff. You decide to tear it up right away. First you tear it in half. After 1 tear the note is in 2 pieces. You put the pieces in a pile and tear them in half again. Now there are 4 pieces. If you do this two more times, how many pieces will you have?

Soccer Schedule

Q: Your mom is in charge of setting up a soccer tournament. She is stuck on how to schedule the games. You decide to help her. There are 8 teams. How many games will there have to be for each team to play each of the other teams?

Your Turn to Turn

Q: You love to jump rope at recess. But you don't like to turn the rope. Most turns jumping last 2 minutes and then the jumper has to take the rope. Recess is 15 minutes long. You want to be sure to get a 2-minute turn jumping, but you want to turn the rope for as little time as possible. How many people can be in line in front of you?

Doggy Boots

Q: You've decided to start a business making snow boots for dogs. You've made a chart to help you figure out how much fabric you need:

SET OF BOOTS	FABRIC
1	0.5 yards
2	1.0 yards
3	1.5 yards
4	2.0 yards

How much fabric will you need for 10 sets of boots?

The Solution

A: There have to be 28 games. You could make a chart to help your mom. You could name the first team "A," the next one "B," and so on:

Team A plays B, C, D, E, F, G, and H. (7 games)

Team B has already played Team A, so:
Team B plays C, D, E, F, G, and H. (6 games)

Team C has already played A and B, so:
Team C plays D, E, F, G, and H. (5 games)

Team D has already played A, B, and C, so:
Team D plays E, F, G, and H. (4 games)

Team E has already played A, B, C, and D, so:
Team E plays F, G, and H. (3 games)

Team F has already played A, B, C, D, and E, so:
Team F plays G and H. (2 games)

Team G has already played A, B, C, D, E, and F, so:
Team G plays H. (1 game)

Team H has already played every team!

7 + 6 + 5 + 4 + 3 + 2 + 1 = 28 games

The Solution

A: 5 yards. Now if it would only start snowing!

The Solution

A: You would have 16 pieces. You threw them in the recycling bin on your way out of class. Hopefully no one has taped them all back together...

The Solution

A: You should be seventh in line. There should be 6 people in front of you, so you will only have to turn the rope for the last minute of recess.

Too Many Rats!

Q: You are dreaming that your pair of mole rats had babies. The babies grew up very quickly and had babies themselves. This kept happening! You had 1 pair of mole rats, then 2 pairs, then 3 pairs, then 5 pairs, then 8, and then 13! In your dream, you noticed a pattern. How many pairs of rats did you have next?

Skateboarding Is Fun?

Q: You are learning to skateboard. If you fall off your board about every 5 minutes, what is a good estimate of how many times you will fall in an hour: more than 30 times, less than 15 times, or about 5 times?

Square Squares

Q: You are building square shelves to hold your collection of square objects. You have a lot of old square wooden boxes, left from your grandfather's collection of square things. If you use 1 box, you can make 1 square shelf. It will take 4 boxes to make a bigger square.

How many boxes will it take to make the next bigger square?

Too Much or Too Little?

Q: Sometimes it's better to overestimate; other times it's better to underestimate. Which do you think is best in each case?

A. The time it takes to get to school on the first day

B. The amount of butterscotch sauce that will fit into your bowl of ice cream

C. The number of hot dogs you can eat without getting sick

D. How far away you need to be after you spray your sister with the hose while she's napping in the sun

The Solution

A: You had 21 pairs. This a Fibonacci number pattern, where you add the last two numbers to get the next one. The pattern continues forever... luckily you woke up before the world was overrun with mole rats!

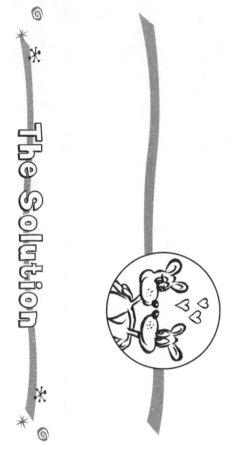

The Solution

A: It will take 9 boxes.

The Solution

A: You would probably fall off about 12 times in an hour, so "less than 15" is the best answer. Ever consider taking up a more relaxing sport...like yoga or badminton?

The Solution

Answer:

A. Overestimate, you wouldn't want to be late for class!
B. Underestimate, unless you like to clean up sticky messes!
C. Underestimate, or you'll have a bad belly ache!
B. Overestimate, unless you can run very fast!

Brain Drain

Q: You weigh 90 pounds. A brain weighs about 3 pounds. There are 26 students in your class. Would the total weight of all of their brains be more or less than your weight?

Piggy Treats and More

Q: You are buying supplies for your pet pig. You need hay at $10.50 a bale, vitamins at $8.95 a bottle, a brush which costs $4.25, and piggy treats at $6.95 a jar. What is a good estimate of much this will all cost, to the nearest whole dollar?

Talking Pigs and Peanuts

Q: You are reading the paper. A headline says "12,500 People Listen to Speech by Talking Pig." Another headline says "245 New Jobs at Peanut Butter Factory." Which headline is probably an estimate? Why?

Vacation!

Q: Your family is taking a camping trip during spring break. You are stopping to eat lunch at your grandparents' house on the way, which is about an hour. It takes about 4 hours to drive to your grandparents' house. If it is about 2½ times as far from your house to the campground as from your house to your grandparents' place, about how long will the whole trip to the campground take?

The Solution

A: Less. The total weight of the brains would be about 78 pounds.

The Solution

A: The number at the speech is probably an estimate since it is hard to count a whole large crowd. By the way, just what kind of newspaper do you read?

The Solution

A: $31 would be a good estimate. That is one pampered pig!

The Solution

A: 11 hours would be a good estimate, since you are stopping for lunch. 2½ x 4 = 10, plus 1 hour for lunch.

Too Much Ice Cream?

Q: You are in charge of ordering ice cream for a school party. There are 400 hundred students at your school. Each will get 1 scoop of ice cream. You experiment with 1 container of ice cream and find that it holds 23 scoops. What is your estimate of how many containers you should buy?

People and Plates

Q: Your parents are hosting a neighborhood party. You are in charge of setting the table. You know that 30 families are coming and each family has 3 or 4 members, but you don't have a more exact count. You decide to put out 90 plates. Is that a good estimate?

Frightening Fingernails

Q: Fingernails grow $\frac{1}{10}$ of an inch a month. You want to grow your fingernails as long as your hair. If your hair is 3 inches long (and you keep it that length), how long will it take to grow your fingernails to the same length?

Top This!

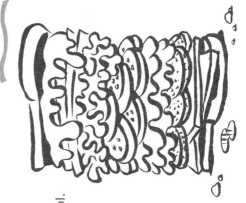

Q: You want to make the biggest sandwich in the world. Or at least the biggest ever seen at your school. You use 1 inch of bread on the bottom, followed by 1 inch of cheese, 2 inches of lettuce, $1\frac{1}{2}$ inches of tomato, $1\frac{1}{4}$ inches of salami, $3\frac{1}{2}$ inches of turkey, $\frac{3}{4}$ inch of mustard, and topped with another inch of bread—yum! How thick is your sandwich?

The Solution

A: No. 90 Plates would only be enough for 30 families with 3 members each (3 x 3 = 90). You know that some families have 4 members, so 105 plates (or slightly more) would be a much better estimate.

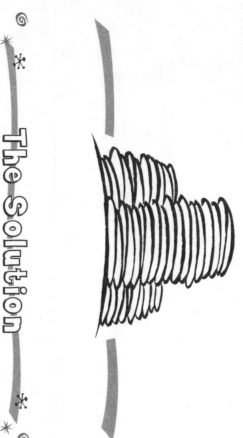

The Solution

A: 12 inches or exactly one foot thick. Now the question is, how will you get that foot into your mouth?

The Solution

A: 18 containers would be a good estimate. (400 ÷ 23 = 17.3) That means 17 would be too little ice cream: You would run out.

You won't have any yourself though. After eating all 23 scoops of your experiment, you won't be craving ice cream for quite some time!

The Solution

A: It will take 30 months or 2½ years to grow your fingernails 3 inches long.

Hop to It

Q: There is frog jumping contest at the State Fair. During the contest, one frog jumps 3 feet and 2 inches. Another jumps 1 yard and 3 inches. The last one jumps 40 inches. Which frog won?

Wacky Weeding

Q: Your job is weeding the garden. You can weed 1 square foot in 10 minutes. The garden is 5 square feet in all. You started at 10:00 a.m. The swimming pool near your home opens at 11:00 a.m. You have to finish weeding before you can go swimming. Can you make it to the pool when it opens?

Tales of a Lizard

Q: You got a lizard for you birthday. At the party, someone accidentally grabbed his tail and it broke off. You found out that it takes 8 months to grow back. If your birthday is March 15th, in which month will your lizard have a complete new tail?

Dr. Frankenstein, I Presume?

Q: A brain weighs 3 pounds. A heart weighs 10 ounces. How many hearts would it take to equal 1 brain?

The Solution

A: The last one. Fortunately for the frog, he jumped right over the river bank and was never seen again.

The Solution

A: In November. I think your lizard better skip your next shindig!

The Solution

A: Yes. It should take you 50 minutes to weed the garden. You will be done at 10:50 a.m., giving you 10 minutes to change and jump in the pool—cannonball!

The Solution

A: Almost five hearts (4.8 to be exact). Now, why exactly did you want to know this?

Choose Your Tool

Q: Your grandfather gave you a strange collection of things for your birthday. In a box are: a long tape measure in inches; a ruler in centimeters; a measuring cup marked in ¼, ½, and ¾ cups; a thermometer; a small scale that can weigh things in grams; and a stop watch. He's asked you to answer some questions and send him the answers. If you get them all right, he's got another surprise for you.

1: Which tool would be best to measure the length of an elephant?

2: What would be most useful in measuring the amount of venom produced by a dozen cobras?

3: What would you use to tell how fast a rhino can charge when he's angry?

4: What would you use to find the weight of a black widow spider?

Mall Moves

Q: You are at the candy store in the mall. You have to meet your mom at the entrance to the mall at 7:30 p.m. It takes 12 minutes to walk to the entrance from the candy store. You want to stop at the music store, a 5-minute walk from the candy store and it is on the way to the door where you have to meet your mom. If it is now 7:10 p.m., will you have time to stop at the music store for 10 minutes, and still be able to meet your mom on time?

Turtle Time

Q: You and some friends are having turtle races. Your turtle, Myrtle, finished the race in 1 hour and 10 minutes. Speedy, another turtle, finished the race in 90 minutes. Flash, a third turtle, finished in 3,900 seconds. Who won the race and by how many minutes?

Lizard Locations

Q: You are trying to find a good location for your new pet green iguana. You have read that they do best at an average temperature of 70 degrees Fahrenheit. You record the temperature in your kitchen, bedroom, and bathroom at different times for a few days:

KITCHEN	BEDROOM	BATHROOM
71	68	73
75	69	70
72	70	68
76	69	69

Which is the best location for your iguana?

A:

1: The tape measure.

2: The measuring cup.

3: The stop watch.

4: The scale.

You are reluctant, however, to send the answers to your grandfather. What if the surprise is the chance to try to measure all of these things for real?

A: No. If you decide to stop at the music store (which is 5 minutes away and stay there for 10 minutes), then walk to the entrance (7 minutes away), you'll arrive at 7:32—2 minutes late. Better buy Mom a CD so she won't be mad!

A: Flash won by 5 minutes. But you missed all the excitement, because you fell asleep waiting for them to cross the finish line.

A: The bathroom. But your dad isn't too happy about it. Who wants to shave with an iguana staring at him?

Stick to It

Q: You are waiting for the bus and it is late. You are bored. You find a pile of 9 sticks that are all the same length. Can you make these into a square? A triangle? A rectangle? (You have to use all of the sticks each time.) Hint: Draw pictures to help you.

Pools for Pets?

Q: You are building a pool for your pet seal. You have 12 square tiles for the bottom. How many ways could you arrange them to form a rectangle? Hint: Draw diagrams to help you.

Sneaky Shapes

Q: You are making up guessing games about shapes with your friends Niki and Pablo. Your guessing game is, "My sides are not all the same length." Niki says, "I have less than 4 sides." Pablo says, "I can be ½ a square." Could you all be thinking of the same shape? If so, what shape would answer all three riddles?

Square into Rectangle

Q: You are sending out invitations to a party. You have printed them yourself on 6-inch squares of bright green paper with purple ink. Alas, the only green envelopes you can find are rectangles that measure 7 inches by 5 inches. Will you have to fold each one to fit it in the envelope? How many times?

The Solution

A: You can make triangle, but not a square or a rectangle. Quick, grab your backpack! You were so busy, you almost missed your bus!

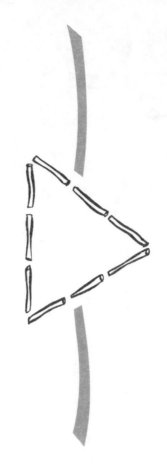

The Solution

A: You could arrange them 3 ways: 3 x 4, 6 x 2, or 12 x 1. Maybe you should have asked your parents though, before you started digging up the lawn!

The Solution

A: Yes, a triangle. Now, make up your own shape riddle and try it out on a friend!

The Solution

A: Yes, once! Now, you better get them in the mail—the party is next week!

Where Are You?

Q: You are playing "Blind Man's Bluff." You take 3 steps north, turn and take 5 steps east, and then turn in the opposite direction and take 10 steps. What direction are you facing now?

Rabbit Geometry

Q: You decide to raise rabbits. You start with a big square pen and 1 rabbit. When you get the next rabbit, you put a fence across the middle of the pen. Now you have 2 places for rabbits. Next you put a fence across the middle of each new pen. You still need more room. You divide each of these pens in half with fences. How many places do you have now?

SUPER CHALLENGE: How many places will you have if you divide the new pens in half? (It may help to draw a diagram.)

Porky, Come Home!

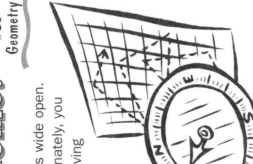

Q: The door of your porcupine's cage is wide open. It looks like he's run away again! Fortunately, you see a trail of fresh porcupine prints leaving his cage. You follow them 10 yards north of the cage, then 3 yards east. Next the prints go 1 yard south and 4 yards west. Finally you follow them 9 yards south and 1 yard east. There's Porky! Where's his final hiding place? Hint: Draw a picture to help you.

Short Cuts

Q: You are cutting out letters for a poster about saving the environment. To save time and make your letters look better, you are folding the paper in half and cutting out the letters on the fold. You start with a "T," and your cut looks like this:

What other capital letters in the word "ENVIRONMENT" can be cut on the fold?

SUPER CHALLENGE:
Are there any letters that you could cut after folding the paper in half twice?

The Solution

A: West. Unfortunately, everyone is to the south and north, so you are still "It."

The Solution

A: Back at his cage! Now, that's a very crafty porcupine!

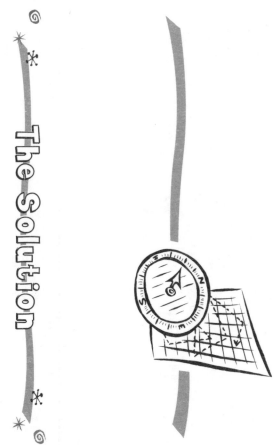

The Solution

A: 8 places.

SUPER CHALLENGE: Now you have 16 places, but the pens are getting smaller...maybe this is enough rabbits for now!

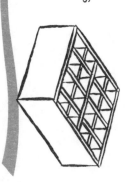

The Solution

A: You can also cut V, I, O, E, and M by folding the paper in half and cutting half the shape.

SUPER CHALLENGE: The letter O can be cut by folding your paper in half twice.

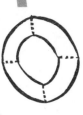

Bad Beets

Q: You don't like beets. You can barely even look at them! If the cafeteria serves beets every third time they serve pizza, what is the percentage of getting beets with your pizza today?

Movie Time

Q: Your family is trying to decide what type of movie to rent from the video store. But everybody is in a different mood! Clever you decide to make a chart of the choices to help make the decision easier. What is the best choice you can make using this information?

PERSON	FIRST CHOICE	SECOND CHOICE
Mom	Adventure	Comedy
Dad	Action	Comedy
Sister	Horror	Adventure
You	Comedy	Adventure
Little Sister	Animation	Comedy

Highly Unlikely

Q: What is the likelihood or probability of these things happening? Rate them using 100%, 75%, 50%, 25%, or 0%.

A. A live giraffe will walk down your street, then lick your face.
B. It will rain in the Amazon rain forest sometime this year.
C. Someone somewhere will have a birthday today.
D. If a woman has a baby, it will be a boy.
E. Snow will fall in Alaska this winter.
F. No one will be born tomorrow.
G. You will live on Mars.

Wiggly Worms

Q: You and a friend decide to play a trick on her brother. You put 2 real worms in a bag with 8 gummy worms. You offer him some worms, and he reaches his hand in the bag and pulls one out. What are the chances of him getting a real worm?

SUPER CHALLENGE:

How many would he have to take out for you to be sure he'd get a real one?

The Solution

A: 1 out of 3, or a 33% chance. Well, is the pizza worth the chance of seeing beets, lying there all red and slippery?

The Solution

A: You decide to pick a comedy because everyone in your family has this as either a first or second choice, except your sister, and you are sort of hoping she'll decide not to watch anyway, so you can get the really comfortable chair!

The Solution

A:
A. 0%...unless you live near a circus!
B. 100%.
C. 100%
D. 50%
E. 100%
F. 0%
G. 0%...unless you are thinking about joining the space program!

The Solution

A: His chance of pulling out a real worm is 2 out of 10 or 1 out of 5, or 20 percent.

SUPER CHALLENGE: In order to make absolutely sure that he gets a real one, you'd have to wait for him to pull out 9 "worms." But your chances of having to run very fast if he does pull out a real worm are close to 5 out of 5, or 100 percent!

Hit or Miss

Q: Your neighbor works every other Saturday. It's Saturday morning and you and a friend are playing baseball in front of his house with no one else is around. You whack the ball with the bat, and it goes right through your neighbor's window! Crash! What are the chances that your neighbor is home and will come running out of the house to scold you?

SUPER CHALLENGE: What would be the chances of him being home if he only worked every third Saturday? Or if he worked two Saturdays in a row and then got the next Saturday off?

Colorful Combos

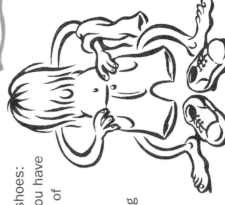

Q: You bought 3 pairs of tennis shoes: 1 orange, 1 purple, and 1 blue. You have a pair of purple shorts and a pair of orange shorts. You have a purple T-shirt and a blue T-shirt. How many outfits can you make from these clothes, if you avoid wearing all the same color at the same time? (You think you'd look silly wearing all purple, for example.) It might help to make a chart.

The Perfect Pet

Q: Your class voted to buy a pet for the classroom. Your group has to make a chart listing what's good and bad about different pets. You want to get the least expensive pet with the most good qualities. Using this chart, what pet should the class buy? What is the second best choice?

Animal	Soft	Smelly	Bites	Hard to find when it escapes	Cheap food	Expensive food	Cost to buy pet	Cost of housing
Hamster	X		X	X	X		Free from Suzy	$20
Rabbit	X	X	?		X		$25	$30
Skunk	X	X			X		$50	$30
Pony	X	X	?			X	$200	$300
Mouse	X	X		X	X		Free from teacher	$15
Rat	X	X		X	X		Free from teacher	$15
Gerbil	X			X	X		$4	$15

Green Means Go!

Q: You are running late for school again. You can make it if the traffic light is green! Here's how the light works: It is red for 75 seconds, green for 15 seconds, and yellow for 90 seconds. (You know it's not smart to cross when the light is yellow and you never do.) When you get to the corner, is it more likely you'll be able to cross or not cross?

The Solution

A: You have a 1 out of 2, or a 50-percent chance of your neighbor being home in this case.

SUPER CHALLENGE:

If he only worked every third Saturday, the chance that your neighbor was home would go up to 2 out of 3, or 66 percent. Your best bet would be if he worked two Saturdays in a row—then you'd only have a 1 out of 3, or a 33 percent of encountering your cross neighbor. But your chance of getting in trouble is still 100 percent!

The Solution

A: You could make 11 outfits from these clothes. That's enough for more than two weeks of school! What a fashion plate!

The Solution

A: The best pet would be the gerbil, since it has only one bad thing (hard to find when it escapes), two good things (soft and easy to get food), and only costs $19 with its cage. The rat is the next best choice, but your teacher will never go for it!

The Solution

A: Neither. The light is green for 90 seconds and yellow or red for a total of 90 seconds. That means your chances of being able to cross is exactly 50%. You're in luck, however, the light just turned green. Now stop reading this and cross the street already!